LIGHTNING AND BOATS

A Manual of Safety and Prevention

LIGHTNING AND BOATS

A Manual of Safety and Prevention

Michael V. Huck, Jr.

Seaworthy Publications

Brookfield, Wisconsin

Published by Seaworthy Publications
17125C West Bluemound Rd., Suite 200
Brookfield, Wisconsin 53008

PRINTED IN THE UNITED STATES OF AMERICA

Illustrated by Tom Baker

Lightning photo on cover courtesy of Steve Hodanish

Front cover photo of static dissipator courtesy of Lightning Master Corp.

Back cover photo of static dissipators courtesy of Island Technology Inc.

ISBN: 0-9639566-0-4

Huck, Michael.
Lightning and boats : a manual of safety and prevention / Michael V. Huck, Jr.
p. cm.
Includes bibliographical references.
ISBN 0-9639566-0-4
1. Boats and boating - - Safety measures. 2. Lightning.
3. Lightning protection. I. Title.
VK200.H83 1995
623.88'8--dc20 95-232
CIP

To Michael and Logan,

~~~

my two bolts of lightning.
~~~

CONTENTS

INTRODUCTION

The purpose of this work is to explain methods of lightning strike protection and prevention for the safety of your boat and crew. As such, I have tried to explain the fundamentals of lightning storms, how they are created, how they can be forecast, and in general, how they behave. Since each boat will usually require a unique protection plan, I have provided the basic theories and guidelines so that you will be able to design a lightning protection system that makes sense for your boat. Finally, I have exposed some of the potential shortfalls that can exist in any boat grounding system. It is my hope that, after reading this book, you will be able to evaluate the protection system on your boat and determine if safety modifications might be needed.

Thunderstorms and lightning have long been the bane of my boating existence. I've spent much time on shore either waiting for a storm to develop, or waiting for a storm to pass. On the other hand, when things have been going badly in a race, I've often listened and hoped for the sound of thunder which would send the fleet scurrying for shelter.

As I grew in boating experience, my visceral fear of thunderstorms never left me. I could sail in heavy weather, I could sail short handed. I could travel, navigate, and generally handle most situations on the water. But, I could not fight a lightning strike.

As my voyages extended beyond the confines of lakes and protected bays, it seemed inevitable that I would encounter lightning in a situation that I could not escape. Then, one day, off the island of Eleuthera, I had the experience of having a storm hurl lightning bolts all around me. I couldn't get out of the way in time, and I couldn't get off the boat. My protection from the power of a lightning strike resided in chance, and an unfounded faith in "being grounded".

The phenomenon of lightning is like many of nature's more flamboyant manifestations. It is surrounded by legend, old wives' tales, and some outright errors. The unpredictable nature of the event has contributed to the sometimes contradictory advice on how to prevent lightning from striking, and how best to minimize injury or damage should a strike occur.

Although most lightning is associated with storms, there is always the possibility of a strike whenever a storm is in the area. The "bolt from the blue" phenomenon occurs when there may be active convection currents causing ionization in your area, but no cloud associated with it. Clear ionization may establish a strong enough charge to attract a strike from a cloud system which seemed too far away to worry about. Fortunately, this phenomenon is a rare occurrence. Most storms only generate strikes in their immediate area.

The destruction that can be done by lightning has been complicated by the introduction of sensitive electronics on the boat. It takes very little current to fry an integrated circuit, and the high cost of most electronic gear raises the stakes in the task of designing an adequate protection system.

In Florida over 200 hundred people die each year from lightning strikes. The fact is, a lightning strike is so cataclysmic that there are no absolutes - except to be somewhere else when it happens. But, there are some definite measures that can be taken for protection and prevention. In fact, a few simple measures may mean the difference between the current from a mild bolt moving easily off the boat, or absolute chaos.

I have a friend who recently had his boat sustain catastrophic damage from a lightning strike. I can't say if the damage could have been prevented by following the procedures outlined in this work, but the chances of the strike happening could have been greatly reduced. Furthermore, while predicting the impact of a major strike is not really possible, directing the secondary current, (or side-flashes), and handling the induced current of the magnetic pulse accompanying a strike usually is possible, and not too difficult to achieve.

The material drawn upon for the information in this book comes from a variety of sources. The procedures discussed have been shown to reduce the destruction visited upon a boat by a direct or indirect lightning strike. It is impossible to say what degree of power will be transferred through a lightning discharge. Some damage may occur no matter what preventive measures are taken. But, with a little planning it is possible to minimize the likelihood of a strike, and provide the highest level of protection for the people and equipment on board.

LIGHTNING

AND

BOATS

A Manual of
Safety
and
Prevention

I

THE FORMATION OF A THUNDERSTORM

Lightning, one of nature's most awesome phenomena is also one of its most common. The phenomena occurs about 100 times per second on earth. In fact, NASA scientists have identified and studied lightning flashes on planets throughout the solar system.

What we observe as a single flash is actually the result of a complex sequence of events. The truth is that lightning is not a single flash at all, but a series of exchanges occurring in a very short time that are only partially visible. The process starts high above the earth.

THE THUNDERSTORM

The thunderstorm begins life as a cumulus cloud. These clouds are the result of warm air rising to a point where the temperature and the reduced atmospheric pressure causes the water vapor to condense and become visible. If the day is dry, or the temperature differences not sufficient, these clouds may stay fairly small, and become what are called "fair weather clouds". If the air is carrying a heavy load of moisture, some of the moisture may fall out of the cloud in the form of rain.

If the upward air currents are strong enough, they will continue carrying the water vapor upward. The reduced atmospheric pressure will allow the cloud to expand, and it may combine with other clouds to form a larger system. As the water vapor is pushed higher, the temperature drops, and the water may freeze and form snow or larger ice crystals. These ice crystals are less dense than water - remember water expands as it freezes - thus allowing the cloud to grow yet higher and wider. These ice crystals form the head of the familiar anvil-shaped thunderhead.

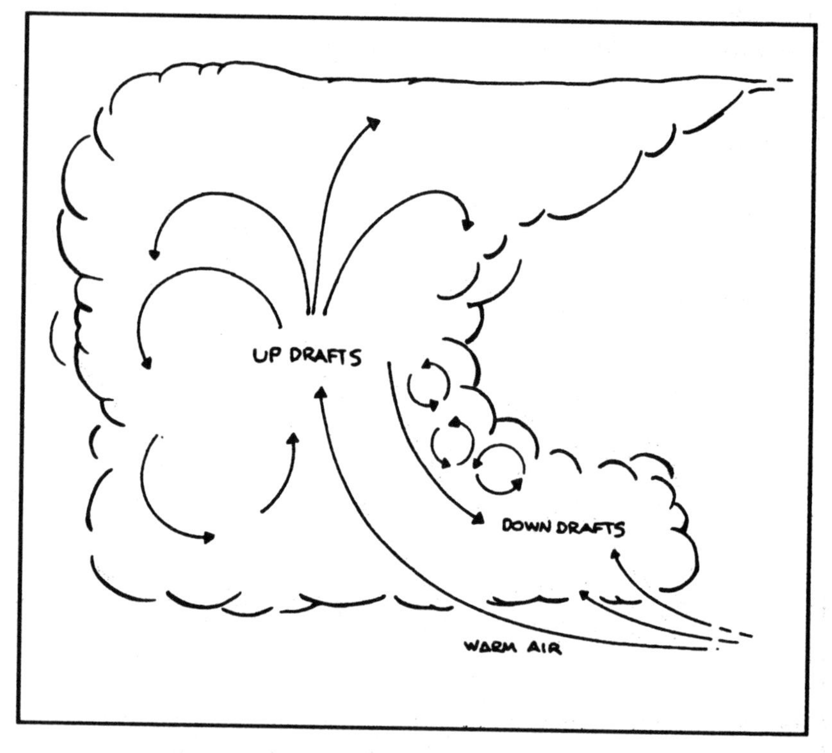

Figure 1. Diagram of air currents in a thunderhead.

As the air mass itself rises, it also cools and eventually starts to fall back through the storm. This air mass encounters other rising air columns and gets moved aside. The cool air may be carried aloft again by yet stronger rising currents. Meanwhile, the water particles, which are called "graupel", will start to fall inside the storm. Particles which have accumulated enough mass will fall to the ground, and particles which are still too light will be sent back up by another rising air mass.

Current theory attributes the establishment of electrically charged areas inside the cloud to the action of these two processes. The cold air falling will "rub" against the warmer air rising and grab some electrons. The energy contained in the warmer air means that the electrons in the molecules are more active and mobile, and more easily shed.

Lightning researchers in Switzerland and South Africa have theorized that the larger water particles take a negative charge, or acquire more electrons after hitting the lighter, rising particles. The result is a concentration of positive charges at the top of the storm, and negative charges in the mid to lower portions.

We can visualize a thunderstorm as a giant storage battery floating through the air; largely positive at the top, largely negative at the bottom. In fact, a single storm carries the potential energy of a small nuclear power plant. But, unlike a storage battery, which is a fairly stable power reservoir, a typical thunderhead has a small region of positive charge at its base. This small, positively charged area at the base of the storm is thought to be key in the formation of lightning.

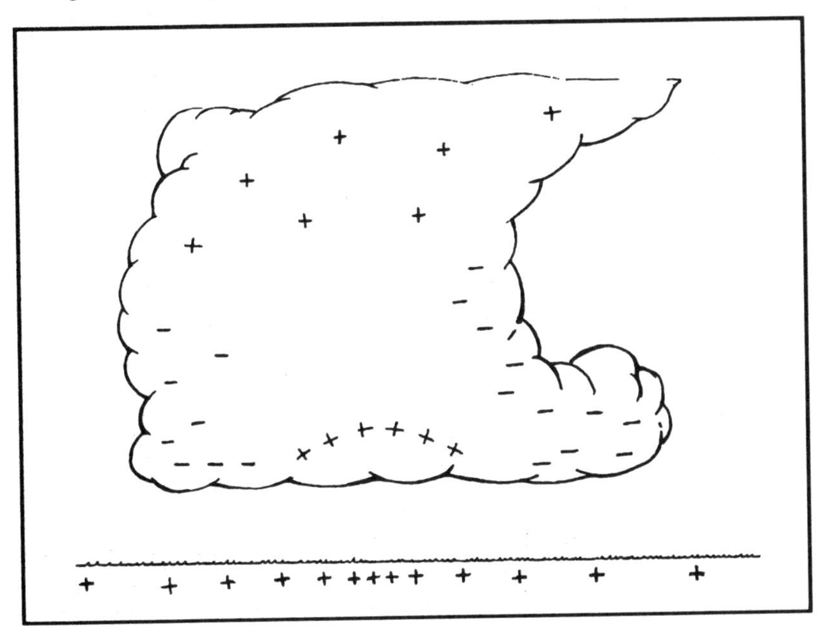

Figure 2. Diagram of the charge distribution inside an active thunderhead.

Once the difference in potential within the cloud becomes great enough, the electric field will overcome the resistance of whatever separates the charges, and there will be a discharge equalizing the potential. This discharge takes the form of a lightning flash. It can occur between two areas in the same cloud, between two separate clouds, between the cloud and the air, or between the cloud and the ground.

LIGHTNING PROPAGATION

The actual events culminating in a lightning flash are still the subjects of some dispute, but for our purposes the basics are defined well enough to give us a good idea of what goes on during the stroke.

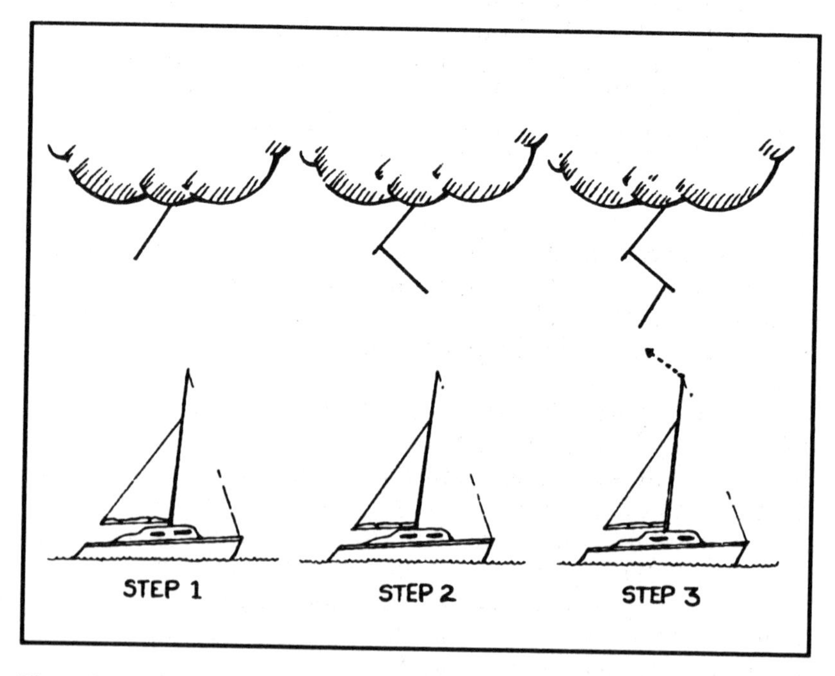

Figure 3. Step 1 - The leader moves downward. Step 2 - The leader branches out at approximately 50 meter intervals. Step 3 - As the stepped leader moves close to ground, a streamer comes up to meet it. When the leader and streamer meet, a return flash will be generated.

As the charged areas within a thunderhead reach a certain potential, the air between them becomes ineffective as an insulator. The charge difference prior to the point of current flow between these areas of the thunderhead is called the electric potential. When the difference in charge between these areas becomes great enough, current will flow across the gap between the areas; a gap that would not normally be able to conduct electricity. The flow of current neutralizes the potential by transferring the excess electrons from the negatively charged side to the positively charged side.

The process begins by the journey of a series of charged columns moving toward an area of opposite polarity. Most often these are negative charges moving toward an area of positive charges. These negative charges move at intervals, or steps of about 50 meters in length and they may branch out at each step. It is the branching which causes the forked appearance of the lightning flash. These charged columns are called leaders. While the leader moves downward establishing its path toward the ground, it appears that the flash follows it from the cloud to the earth, but this is not actually the case.

As the leader approaches the area of opposite charge, the charge becomes concentrated. Leader-like structures called streamers may be sent from the area of opposite charge. Once the streamers meet the leader, a return flash is generated. This flash is the visible part of the lightning strike.

The visible part of the stroke is actually made up of several flashes. Each flash after the first is the product of a leader from the cloud followed by a return flash from the ground. All of these flashes occur over the course of several micro- seconds, and are not individually visible to the eye. As many as 48 of these leader-flash pairs have been detected by scientists in a single lightning strike. The bolt appears to be pulsing slightly when this happens. Longer lasting strikes carry more current with them, and constitute a greater hazard.

The discharge process initiating the lightning strike starts as a breakdown in the lower part of the cloud. Containing the large negative charge and the smaller positive charge, a discharge in this lower area breaks loose the electric potential and initiates the leader. The leader then travels toward an area of opposite potential on the ground, or toward another cloud, or toward the positive upper area within the same cloud.

As the thunderstorm builds in intensity, it also starts to concentrate a positive charge on the earth below it. The earth normally has a charge relative to the atmosphere. As the thunderstorm generates its internal electric fields, the earth will develop a more intense, localized positive charge. The locus of this charge moves underneath the storm as it travels in the air currents. The closer a point on the ground is to the internal fields of the storm, the more concentrated the charge upon that point will be.

Visualize a carpet of net positive charge unrolling underneath the storm. As this carpet covers the surface, it follows the terrain beneath it. The carpet lays over objects; the higher the object, the closer it is to the cloud and the greater is the field strength surrounding it. The fact that things which are tall are generally narrow, for example a broadcast tower or a sailboat mast, concentrates the potential. The smaller the area available to hold the charge, the greater the concentration of electric potential will be. When there is a stronger field located on a small area, the concentrated field becomes an area of potential lightning attraction because the area is highly charged, the charge is focused, and the field is closer to the opposite charge contained by the cloud. Such an electrical field will attract the leaders descending from the cloud.

Remember we are talking about relative strengths here. The ends of blades of grass will be areas of greater charge concentration than a flat surface. A tree would have a greater field strength than the grass etc. The object or area with the greatest field strength will generally attract the strike, which has obvious implications for sailboat masts, upraised golf clubs, and similar items.

THE BIRTH OF A THUNDERHEAD

It is a warm summer day in Florida. As the day heats up, the air rises and cools. The water vapor carried by the air condenses and forms fluffy white cumulus clouds. Although the water vapor is now visible, it is still supported by the rising air. More and more water condenses out of the air as the air rises and the atmospheric pressure and temperature are reduced. Some of the air cools to the point where it tumbles out of the rising column and starts downward. The falling air rubs against the rising currents and may encounter even warmer air with enough energy to start it back upward again. This warmer air brings more moisture with it, and the cloud continues to grow.

As the air currents rub against each other, electrons transfer from one column to another. Water droplets also form and begin to fall through the cloud. The larger droplets attract loose electrons from other lighter droplets which are still rising. The heavier droplets fall to the lower portion of the cloud, and the cloud starts to build areas of distinct charges.

As the breeze from the ocean starts to flow inland to replace the warm air that has risen with the cooler moisture-laden air, the cloud expands further. The new air gets heated by the warmer land then rises into the cloud. Then, the cloud builds further in height. When the cloud builds high enough, the water in the top freezes, and the cloud expands even more. Eventually, the cloud expands to reach the upper winds and starts to move to the east.

Meanwhile, as the charge in the cloud builds and concentrates into positively and negatively charged areas, a shadow of positive charge follows the cloud along the ground. This charge lays over objects on the ground, and intensifies when it gets closer to the negative charge carried in the lower part of the cloud. A sudden disturbance in the base of the cloud between the small area of positively charged ions at the bottom and the negative charge slightly higher releases a flood of electrons, (called a leader), which starts downward toward the positive potential on the ground. As the electrons travel downward in steps, the positive charge on the ground intensifies.

NASA PHOTO

Figure 4. An electrically active storm.

The positive charge on the ground concentrates in areas closest to the stepped leader. When the field reaches a threshold strength, ions start traveling from the positively charged point on the earth, forming a streamer, and meet the leader coming down from the cloud. The electric circuit is then complete.

The trail of ions provides a path that is more conductive than the air, and a lightning flash emanates from the ground. It follows the path of the leader back to the cloud. The difference in electric potential is reduced. But, the first flash almost never exhausts the potential difference, and now a ready made ionized path exists for the development of more flashes. Another leader travels down from the cloud, and another return flash is generated from the ground. This sequence may repeat itself as few as 3-4 times, but it has been shown to repeat as many as 48 times. Eventually the potential is reduced to the point where it cannot sustain the channel, and the lightning ceases.

II

CHARACTERISTICS OF LIGHTNING

A bolt of lightning is one of nature's most powerful phenomena. The leader-return flash combination has broken down the resistance of a channel 3 to 14 kilometers long. The current generated is more than enough to light a small city. Frequently as much as 50,000 coulombs are involved in an individual flash. A coulomb is a current of one ampere per second, and the lightning flash may occur in a few tenths of a second. The temperature reached at the core of a lightning flash exceeds 30,000 degrees Kelvin, which is several times hotter than the surface of the sun.

The core of the stroke represents the path of the current. As the current moves through the air, the air around it becomes luminous. As the current decreases, the luminosity diminishes, and finally the current stops passing through the core. The core diameter is very small compared to the visible size of the flash. Experiments placing a fiberglass shield over a lightning terminal show that the holes burned through the shield are tiny compared with the observed size of the strike.

Notwithstanding, the small size of the core, and the short period of its existence, a strike still has amazing destructive power. Evidence of the power of lightning can never be forgotten by anyone who has seen a splintered tree, or watched the cable running from a lightning rod glow after being struck.

Lightning is not a homogenous event. Strikes vary considerably. One strike may deliver over one thousand times more current than another. Strikes can last as long as ten seconds, or be as brief as two tenths of a second. Strikes that are longer lasting are capable or transferring more of the heat in the core of the lightning bolt to the object struck. Consequently, longer lasting strikes can be far more damaging. It is virtually impossible to predict what kind of strike will occur at any given time.

In addition, there are two major types of strikes. The most common carries a negative charge from the cloud to the object struck. About ten percent of strikes carry a positive charge. These positive strikes transfer about 5 times more current than their negative counterparts. Usually there will be one positive strike in each storm.

Lightning takes place by bridging areas of opposite charge. These opposite charges can be within a single cloud, between two clouds, or between the cloud and the air. 80% of all lightning occurs within the same cloud. The remaining 20% will be split between the other clouds, the air, and objects on the ground.

The damage caused by lightning usually is a factor of heat transfer. Either the flash itself lasts long enough to pass along a significant portion of its temperature to the object struck, or the current flowing through the object causes a high temperature as it overcomes the resistance in the object. The heat generated by lightning can do strange things; it can vaporize water, melt metal objects, even cause explosive expansion.

On the other hand, a short duration strike transferring a relatively small charge may simply pass through an object leaving it basically unchanged. Anecdotal evidence abounds of people and objects taking the full force of a lightning flash and remaining essentially unharmed.

Lightning is also responsible for some indirect effects as well. The indirect effects of a lightning strike can be just as destructive as the strike itself, particularly to sensitive items such as modern electronics. As lightning neutralizes the potential difference between its source and target, the area surrounding the target remains charged. The charged area will neutralize itself, and current will flow toward the target. This event is known as a side-flash. The magnitude of this current is not the same as the strike itself, but still will be more than sensitive electronic circuits can tolerate.

A lightning strike will also generate an electromagnetic pulse which can induce current in conductors in the area. A portion of this pulse is heard as static on your AM radio. The electromagnetic pulse contains wavelengths of both the visible and invisible spectrum. The magnetic portion of this pulse rises and falls rapidly . As such, the pulse is capable of creating current in electrical equipment nearby. An electromagnetic pulse from a lightning strike could devastate the solid state circuitry in a boat's electronic gear.

Effective protection from the effects of lightning is difficult to achieve. If the strike is one of the high powered, long duration variety, protection may be impossible to achieve. Fortunately, statistics indicate that the likelihood of being hit by the latter (positive) type of strike is much smaller than the chance of being hit at all. Therefore, for the boater, the objective must be to design a system that will significantly increase the vessel's chances of sustaining a strike without injury to the people on board or major damage to the vessel and it's components.

In the next chapter we will discuss how to anticipate the onset of lightning, and how to take the most important step of safeguarding the lives of the people aboard.

III

PROTECTING YOURSELF ON THE WATER

Although the incidence of strikes occurring on the water is quite low, being out in the middle of a storm is not for the faint hearted. Since lightning is usually accompanied by high winds, strong gusts, and driving rain, arranging to be somewhere else when a storm happens is always the best precaution. At times when the formation of storms is likely, a knowledge and understanding of the observable atmospheric changes may allow you time to reach shelter.

One of the causes of electrical activity is the friction generated by columns of air moving upward and rubbing against stationary colder air. These columns of heated air are called thermals. Conditions that contribute to the formation of these thermals could also lead to the formation of thunderstorms. The thermals will initially form puffy clouds known as cumulus clouds when the moisture in the air condenses as the air cools. These cumulus clouds are usually known as "fair-weather" clouds. They have not reached the point where they represent anything more dangerous than a temporary interruption of the sun.

But, these clouds can yield a great deal of information about conditions later in the day. If the cumulus clouds form early in the morning, the thermals creating them will eventually get more powerful. The cumulus clouds will grow, combine, and form the cells which will eventually become thunderstorms.

Gauging the speed of their formation is a useful predictive tool for the boater. The faster the clouds grow, the more likely the clouds will become electrically active.

The sudden appearance of puff-clouds means nothing. Anytime air reaches the temperature needed to condense moisture, these puff-clouds can form rapidly. But, If you look at the tops of the clouds with a pair of polarized sun glasses, and binoculars, you can judge the strength of the underlying thermals.

In Chapter 1 we learned that as a storm cell builds, moisture is propelled upward by rising air currents. Eventually this moisture reaches higher altitudes whereupon it freezes and rises even higher. Finally it reaches high altitude air currents and is pushed eastward, thus forming the familiar anvil shape of the thunderhead. If clouds with these ice tops (or thunderheads) are visible on your horizon, then chances are that clouds nearby your vessel, (clouds whose tops may not be visible), are also forming thunderheads.

Next, the bottoms of the formerly puffy white clouds will turn dark. The growing density of the cloud is from the increase in moisture levels as new vapor is brought up by more warm air currents traveling upward. More moisture means larger particles of water, and more sunlight blocked from reaching the bottom of the cloud.

The important consideration is how quickly this sequence of formation is taking place. The faster the clouds build, the more violent the resulting storm is likely to be. A rapidly building storm will also move more quickly, reducing the time available to seek shelter.

FRONTAL SYSTEMS

Storms harboring electrical activity may also be the result of larger weather systems. A cold front may have a line of thunderstorms at its leading edge. As the front sweeps through, these storms will roll over you bringing high winds, lightning and a temperature drop as the cold air mass moves in.

The severity of the conditions resulting from the passage of a cold front can be gauged by the difference in temperature on either side of the front. If the air mass driving the front is much colder than the air it is displacing, the division between the two will be very sharp. As the colder air slides underneath the warmer air, the colder air will force the warm air higher and cause the moisture within the warm air to condense. This process will lead to friction between the two air masses. The friction will in turn generate electrical activity in the same way that the thermal process does. Eventually, the clouds in the front will develop charged areas and there will be lightning as the potential seeks to neutralize itself.

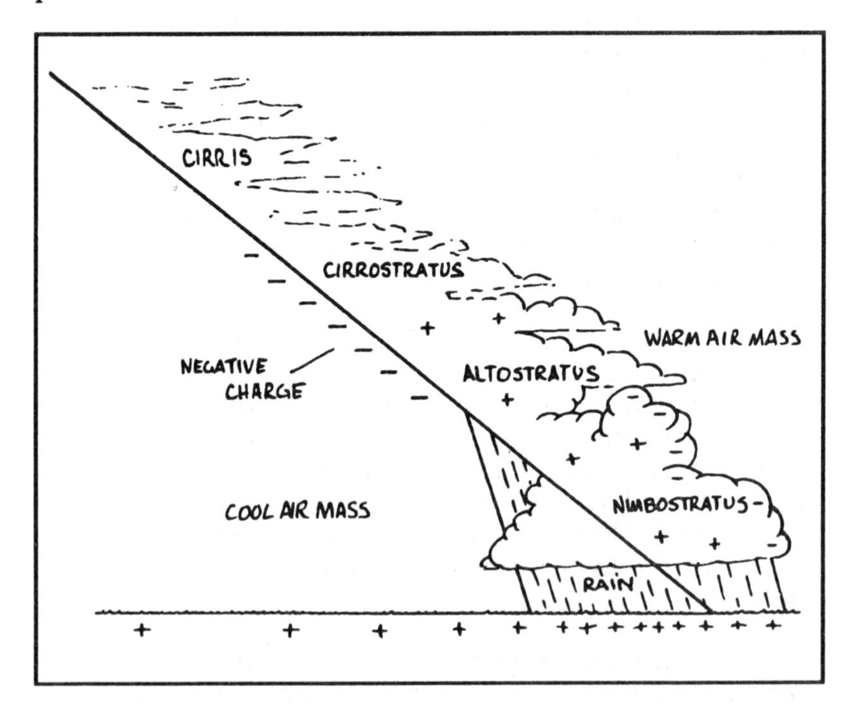

Figure 5. Typical frontal system. The warm air rises over the cooler air, creating friction which causes charges to build up on either side of the front. Note the concentration of positive charge on the ground.

Storms at the forward boundary of the front will form along the leading edge. Since a cold front can extend across an entire continent, the area of storm formation can be quite large. These storms that line up at the edge of a front are called squall lines. A squall line is a smaller area within the frontal system that spawns locally high winds and hard rain. It is possible to have a squall line without any electrical activity, but the violence of the squall itself should still be avoided when possible.

Cold fronts are usually quite well defined and are easily forecast. The temperature changes and cloud patterns give the meteorologist plenty of data to develop the projected track and speed of the system. But, the path of the front can be altered by a number of factors. For examplc, the front may encounter a locally stronger high pressure area near the trailing edge of the opposing air mass which might force the front further north than projected. Or, the low pressure area which usually drives the system may intensify and speed the process along. In any event, it is usually sufficient for the mariner to know that frontal passage may occur, and then to keep his weather eye out for signs of its approach.

There will be some high clouds preceding the actual front, but the distance between them and the more serious weather to come varies a great deal. After being alerted by the higher clouds, a close eye on the horizon will pick up the lower clouds marking the boundary of the air mass. If the system has formed a squall line, the characteristic roller cloud will be seen. This cloud will move very quickly, and it is important to develop plans to seek shelter. As the squall passes over, visibility will drop dramatically. Strong rain and wind will make it difficult to navigate, and the commotion can distract even the most experienced pilot. A predetermined course to a sheltered location will make it easier to concentrate on simply piloting the boat.

By contrast, a warm front may also develop thunderstorms; but, their approach is announced well in advance by the onset of high cirrus clouds. The warmer air behind the front advances over the colder air. At the front there is still the interaction between the cold and warm air masses which causes friction between the two. Similarly, the friction develops electrically charged areas, and eventually, thunderstorms.

DETECTING POTENTIAL LIGHTNING DANGER

The various types of potential thunderstorms manifest themselves in entirely different ways. For example, a typical summer thermal storm may build all afternoon, then suddenly start moving. It may be impossible to tell when the cloud changes from being a feature of a nice afternoon, to a threat. But, what we can depend on is that as the cumulus cloud develops, intra-cloud discharges will take place. These discharges generate thunder. This thunder provides two indications to the boater; it alerts you that a cloud is electrically active and will be a threat if the cloud moves over your area; and the thunder can give you an indication of how far away the storm is.

Clouds are somewhat similar to mountains. They rise up without much else for visual reference. The lack of reference makes it really difficult to accurately gauge their distance from you. When thunder begins to be heard, a simple calculation will tell you how far away the storm is and give you a time frame in which to plan your trip to shelter. The farther you are from a place of safety, the sooner you need to think about heading there.

The unpredictable nature of these storms may result in them simply passing by with no discernible effect. In Florida, there would be no boating in the summer if the sight of building cumulus clouds were enough to send everyone to shelter. They will build up during the early part of the day, and suddenly start moving in the afternoon. All you can do is keep the clouds under observation, and exercise some prudence. The isolated, convection type storms usually have a sharply defined rain line. This rain line will help you track the progress of the storm.

Thunder is simply the sound created by the rapid expansion of the air from the heat created by the current of the lightning strike. This expansion creates shock waves which travel through the atmosphere at the speed of sound, about 750 mph. Since the flash from a lightning strike travels at the speed of light, the difference between the time of the flash, and the time you hear the thunder can give you a idea of how far the electrical activity is from your position.

Thunder travels approximately 2 tenths of a mile per second. After seeing the lightning flash, and timing the difference in seconds between that flash and the thunder, divide the result by 5 to arrive at the number of statute miles between you and the storm. You may be able to find out how fast the storm itself is moving from a local weather station. These two pieces of information will tell you approximately how long you may have to get to a sheltered location.

A problem with this method is tying the thunder together with the actual flash that caused it. Intra-cloud lightning is not always visible. A strike seen in a storm fairly distant may be placed together with thunder from a closer storm. The distance obtained from this pairing would mean nothing. As a rule of thumb, thunder will not be heard from a strike over 25 miles away. If you get a figure like that, keep alert for another opportunity to time another flash-thunder pair.

OTHER STORM DANGERS

Apart from the lightning, the thunderstorm presents a number of other hazards for the boater. As discussed, the formation of the storm itself depends largely on the movement of air masses. The center of the storm also develops a localized low pressure area. This adds a surface component to the movement of the air.

As the storm approaches, the air surrounding it will revolve around the center of the low, and can get quite gusty. The low itself will cause locally windy conditions that can be quite disconcerting. As the storm passes, the wind direction will shift around the low. When there is also heavy rain, visibility can be reduced to a few yards. If the wind direction is being used to orient the heading of the boat, the shifting wind will throw things completely out of whack. With the reduced visibility, it will be possible to get seriously lost. If this takes place in a hazardous seaway, maintaining a good heading will take a good deal of concentration. Breaking this concentration to try and plot a new course to safety will make a hazardous situation worse.

PREPARING FOR A THUNDERSTORM

Racing sailors have a rule of thumb for finding wind: sail for the black clouds. Sometimes the black clouds sail for you. When this is the case, anticipation and preparation can make the difference between waiting out a bit of momentary rough weather, and flailing around with gear, sails, and rigging scattered all about. Anticipation doesn't require that you reef down at the first rumble of thunder. All it means is that the various procedures for storm preparation are discussed. The foul weather gear may be brought up so that all hands aren't down below trying to get into their gear as the storm arrives. Loose items can be secured, and the deck area made ready for work should conditions warrant.

Should you find yourself underneath the storm, the safety of the personnel on board becomes the critical issue. Sending people below is the first element of proper procedure. (There probably won't be many objectors.) The next most important consideration is to limit the metallic objects that anyone on deck contacts. The reason for this is quite simple. Beyond the current imparted by the strike itself, the secondary current flows can be a significant hazard. As we shall see, current flows from the extremities of the boat toward the area struck by lightning. This current takes the path of least resistance. If a crew member is the bridge between two metal items on deck, he may well become that path.

Consider the helmsman leaning on the aft rail and holding the wheel in his other hand. Also, the crew member holding the boom out while leaning on a stanchion is at risk. Electricity may pass through the crew member even if the two metallic items are bonded and grounded. His body may be the path of least resistance at that point.

A positive note is that the steps taken to lessen the risk of current flowing through people, also serve to protect the physical structure of the boat from damage. But, the danger to people from lightning strikes at sea, while still the major concern, is only part of the potential damage caused by the lightning event.

IV

LIGHTNING DAMAGE AND PROTECTION

Nowhere is the capricious nature of lightning more evident than when viewing the damage done by a strike. Some strikes will leave no visible evidence of the current's passage. Other strikes will destroy completely whatever the bolt has hit. The variations can be explained by understanding the lightning itself. As previously stated, each strike is made up of multiple flashes of lightning. Each flash is preceded by a leader. Each flash's energy changes as it draws upon a different location and field strength of the charged area in the cloud. If the conditions are correct, there may be a great number of leader-flash pairs within the strike, and a great deal of electrical potential exchanged.

The current in each flash varies, but the key element in the damage caused by a strike is the duration of the current. In an event measured in milliseconds, little heat can build up to cause damage. But, in the course of several flashes, or a number of current flows, residual heat will build causing significant damage. The current flow must last for a period of time for the heat to build up. As strong as the current in a flash is, the duration of the current in a single flash usually cannot create a high enough temperature to cause damage. Multiple flashes may still not last long enough to cause any damage from heat. Unfortunately, it is impossible to predict the point at which damage from heat build-up will occur. Some strikes have long enough duration to cause major burns. Indeed, some strikes will ignite forest fires.

Heat is a by-product of current flow. In rough terms, the greater the resistance that the electricity passes through, or over, the greater the energy thrown off as heat. When the current from a lightning flash passes through something which does not normally conduct electricity - something which has great natural resistance - heat is generated which can cause the burns often produced by a lightning strike.

For example, current passing from a chain plate to the water through the hull can cause a burn down the side of the boat. As another example, I once saw a boat that was hit by lightning while covered. All of the zippers in the cover were fused together.

Another contribution to the unpredictable nature of lightning damage is the existence of the two types of lightning strikes. A positive lightning strike contains about 5 times as much energy as a negative strike and has a longer duration. These positive strikes are rare, only about 10% of all strikes fall into the positive category. But still, this means that most storms will contain at least one positive strike. As we discussed in a previous chapter, the combination of greater current with longer duration is a destructive mix.

PLANNING FOR PROTECTION

The analogy that links electrical current flow to the water in a river is a useful one. Like the water, the charge will move through the path of least resistance. In order to control the flow of current in a lightning strike, we need to provide this path of least resistance. Resistance is a common electrical quantity, but what we need to consider is the flip side of resistance, or conductivity.

Conductivity refers to a substance's ability to transport an electrical charge. Materials which are highly conductive generally have electrons that are easily shed, and therefore very mobile. These electrons will move, and current will "flow" very easily. Copper is an excellent conductor, as are silver, gold, and most other metals. By contrast, materials that do not conduct electricity easily - dielectrics - have relatively strong bonds between atoms and few electrons available to move. The galvanic table of metals is an example of one type of ranking by conductivity.

But, the fact is, an electric potential that is sufficiently strong will cause even the most reluctant electron to move. Any insulator, be it plastic, glass, or paper, will break down and conduct current.

We need to develop a three-tier approach to protecting ourselves from a lightning strike. First, we need to protect the structural integrity of our boat by providing a path for the major current to travel through, and off, the boat. The current is going to get off somehow. Controlling the exit point can keep the current from building up heat in the first place. Damage caused directly by a strike is not the most common form of lightning problem, but it is certainly the most dramatic. As the current moves down the rigging, it has sufficient energy to travel the most direct path to a ground.

This direct path to ground is the cause of the burn marks on many boats hit by lightning. The current flows through the boat in a narrow path. The resistance of the wood or fiberglass creates heat as the electricity flows through, and the surrounding material is damaged. This kind of damage can be particularly crippling in a fiberglass boat. The laminate typically has microscopic voids that eventually fill with water. This water is vaporized by the heat of the strike and expands. The expansion inside the laminate fractures it along the path of current flow. The damaged area then acts as a hinge. As the boat works, this area will act as a pivot for the hull. Continued flexing around this point will cause fatigue failure.

This first aspect of protection, catastrophic strike protection, is where most systems stop. Most lightning protection systems ignore the other electrical effects of lightning which do not even require a direct strike to manifest themselves.

The second step in a good lightning protection system should be protection from the secondary current flows associated with a strike. Preventing these current flows requires a way to equalize the ground potential for all the metal objects aboard. With a common ground, a difference in electrical potential between, say, the aft rail and the helm should not develop. Since these changes in potential take place at a significant fraction of the speed of light, the electrical fields equalize very quickly. If we can provide a good conductive pathway to a common ground, the current may not attempt to travel through things like people, or fiberglass.

Coinciding with this second area of protection is the bonding of the electrical systems on the vessel. The introduction of sophisticated electronics in the marine environment provides a major opportunity for damage through lightning. This equipment need not even bear the brunt of a direct strike to be affected. One of the characteristics of solid state, microprocessor based instruments is that they require very little in the way of current to do their job. The power requirements of today's electronics are dramatically lower than older equipment that contained vacuum tubes. Consequently, their ability to withstand a surge of high power, from whatever source, is reduced.

Therefore, solid state equipment is extremely vulnerable to the secondary effects of a lightning strike. Indeed, even a strike which is merely "close" may cause enough ground potential difference within the boat's electronics system to scramble the microprocessors. The current may come from wires leading from sensors; it may come from the power supply; or it could be caused by an electrical potential difference that develops in the chassis holding the instruments. Any residual magnetic field left by the current flow through the system will affect the functioning of these sensitive instruments. Bonding the power supply, data lines, and chassis will help prevent discharge.

The third area of protection consists of deterring the strike in the first place. As we've discussed, there is a sequence of events leading to the return flash of the lightning strike. Once the stepped leaders reach the area where a return streamer can be sent from the object about to be struck, the greatest need becomes preventing the return streamer from forming.

V

LIGHTNING PROTECTION SYSTEMS

In theory, a boat at sea would appear to be extremely vulnerable to a lightning strike. We know that lightning is often attracted to the highest point around. We've seen how the thunderstorm draws a ground charge underneath it. Offshore, the mast of a boat is clearly the highest point, and the top of the mast provides a nice place for a concentration of ions. Yet, I have rarely heard of, and I have never been aboard, a boat that has been hit by lightning while at sea.

My theory is that the risk for boats at sea may be reduced by the motion of the masthead, or other high points on a boat in motion. I believe that the movement of a boat through water allows a vessel to more easily shed electrical potential. Since the electric potential is constantly being shed, it rarely builds to the level necessary to generate a return streamer.

By contrast, I have frequently heard tales of boats being zapped at their moorings, or on trailers. A boat in a stationary position exhibits all the characteristics of a lightning attractor. A masthead is full of sharp edges and points which can provide areas of high charge concentration and contribute to the formation of the streamers which are the precursors of a strike. And, since there is no movement through the water, the potential builds without being shed.

Bruce Kaiser, President of the Lightning Master Corporation in Clearwater, Florida puts it this way, "Imagine turning your boat upside down and dipping it in syrup. When you lift it from the syrup, the points from which the syrup drips are analogous to the ground charge accumulation points." The traditional way to handle this problem of charge accumulation is to mount a rod which is electrically attractive and have it take the strike. Then the energy can be directed down a safe, highly conductive path to a ground. The method is to create the ideal conditions for the formation of the ground streamer so that any potential strike will occur at the point specifically designed to handle it.

Once the strike is triggered, the path to the ground is clearly defined, and the energy safely controlled. This method works best with buildings because they have plenty of space for mounting rods, and grounding cables. But, the area at the masthead on most sailboats is already crowded with wind sensors, antennas and lights; any one of which may also build up enough potential to attract the strike.

Furthermore, even with a system for handling a strike, the vessel will still be subjected to all of the secondary effects of the strike. Even if a strike passes safely through the boat and out to a ground, the current passing through the ground wire will generate an electromagnetic pulse which will induce current in all of the conducting materials on the boat. The neutralization of the potential difference between a cloud and an object struck will also lead to a secondary current flow (or side-flash). This side-flash will occur between points with a different ground potential than the object taking the strike. Therefore, the best approach is to develop a system for strike avoidance; a system designed not to attract and control a strike, but rather to limit the likelihood of a strike occurring at all.

Actually, a system designed for lightning attraction could function well to protect a larger number of boats in a harbor, or in a parking lot by attracting the strike to a terminal expressly set up for that purpose. An installation could be designed to intercept, or attract, all potential strikes which might be in the area. NASA uses this technique to protect its spacecraft while on the launch pad. A mast is mounted on the launch tower, and a catenary wire is led from the base of the mast to a ground several hundred yards away from the pad. The wire directs the current from a strike away from the sensitive instruments that are on the launch vehicle, and allows that current to dissipate harmlessly at some distance. A similar system could protect a large storage area, or even a small harbor.

There are certainly limits to the radius of protection offered by this system, but it could easily be adapted to serve an area the size of most marinas. The NASA system is relatively simple. There are however, some design criteria which must be met if the rod is going to be an effective streamer emitter. Remember that the final phase of the lightning strike is the growth of streamers emanating from the ground to meet the stepped leaders coming down from the cloud.

An effective lightning rod must be sharp; not just physically sharp, but electrically sharp. The intensity of the electric field depends on the radius of the point collecting the charge. A smaller point will generate a more intense field. An intense field will in turn more efficiently retard streamers until they are powerful enough to cause a strike.

There is one problem with this system. Even the sharpest rod will become dull after prolonged exposure to the elements. The rod certainly won't appear too dull from the ground, but the attractive forces we are dealing with are so small, that minor degradation in the system will affect it. The problem can be addressed by placing a narrow gauge wire at the end of the rod. The point cannot degrade any further than the radius of the wire. Unfortunately, a narrow wire will also have a fair amount of resistance, and will melt from the heat generated by the passage of current through it. After the first strike, the wire will lose most of its lightning attracting ability.

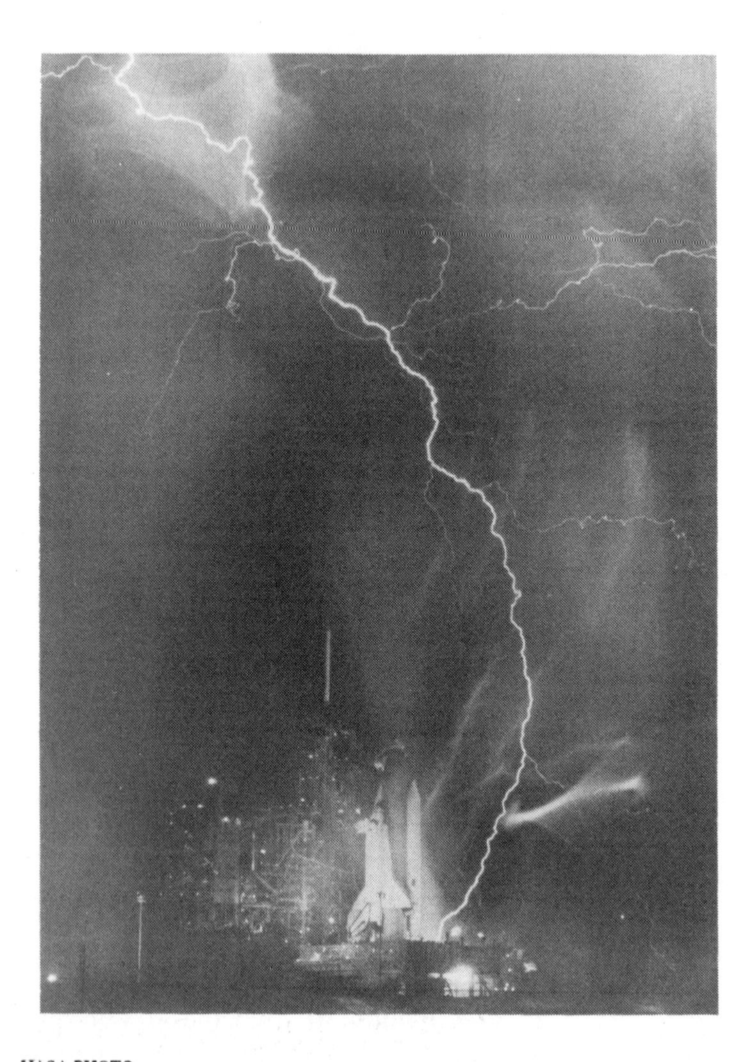

NASA PHOTO

Figure 6, 7, and 8. Powerful lightning strikes the launch pad at Kennedy Space Center, Florida. Scary but survivable, with the right equipment. A catenary wire carries the potential of the ground at the remote end, protecting the sensitive instruments onboard the space craft.

NASA PHOTO

Figures 7 (above) and 8 (below). Powerful Lightning strikes the launch pad at Kennedy Space Center, Florida.

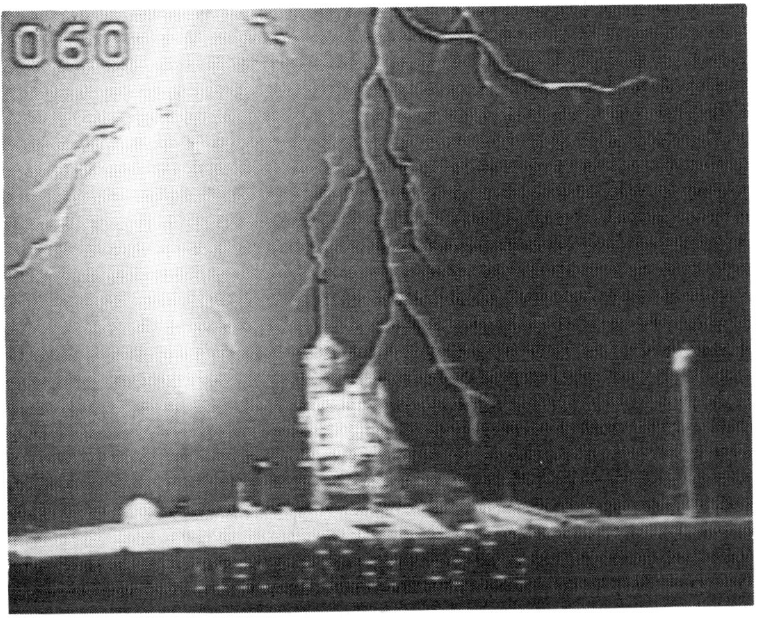

NASA PHOTO

The mechanics of lightning protection on a boat are somewhat different than for a system designed to intercept a strike. On a boat, the actual strike is the worst case scenario. We do not want to invite the current aboard. But should a strike occur, we need to give the current a path through and off the boat in the most direct route possible.

Let's start with the top of a mast. As a cloud pulls the ground charge along underneath it, the charge collects on whatever is closest to the field in the cloud. The field intensity increases with the charge density, and the charge density increases with a decrease in the area holding it. As we know, the top of a sailboat mast, and even the mast of a powerboat is crowded with antennas, lights, and wind sensors. Most of these instruments are quite narrow in diameter, and provide ample opportunity for a build up of sufficient static charge to emit streamers upward. Keep in mind that no matter how well grounded a boat is, the objective is to deter the strike. The way we can protect ourselves is to encourage the static charge on our vessel to dissipate before it reaches the necessary potential to generate streamers.

Our protection again comes from the operation of the point discharge principle. As previously stated, a sharp lightning rod will develop a stronger field and shed electrons more easily than a dull rod. This sharpness allows a rod to retard the streamers until they can attract the lightning. We can take advantage of this predictable behavior by creating a multitude of points which will all easily shed electrons to dissipate a charge before it reaches the potential to send a streamer upward.

A static dissipator differs from a rod by having a large number of electrically sharp points in a relatively small area. This prevents the charge that is being shed from building excessively on any one point. If the charge is kept from building, it will not be able to form a streamer extensive enough to attract the stepped leaders approaching from above. In other words, the streamer which may emanate from the top of your mast is competing with streamers from any other objects or surfaces underneath the storm. As the stepped leaders come down from the cloud, they are attracted to the higher field strengths.

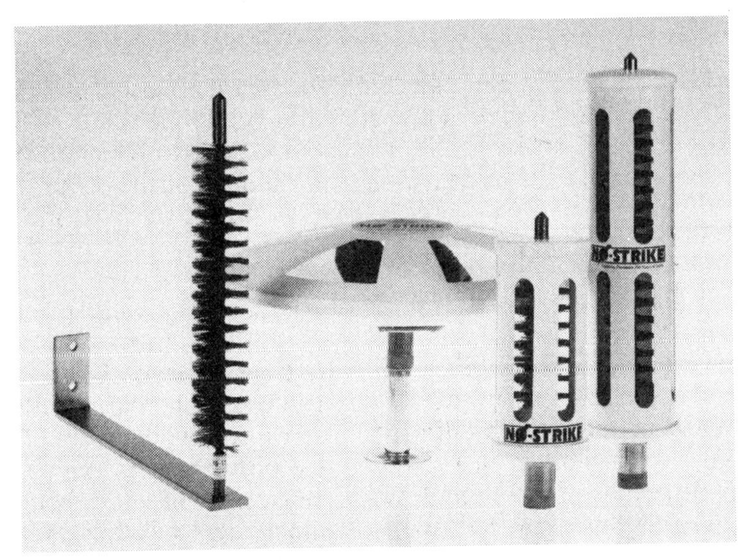

COURTESY OF ISLAND TECHNOLOGY, INC., CHERRY HILL, N.J.

COURTESY OF LIGHTNING MASTER CORP., CLEARWATER, FL.

Figure 9. The static charge from a vessel can be bled into the atmosphere through the mass of small points on a static dissipator. By using a static dissipator, the magnetic field strength on the vessel is greatly reduced making its potential similar to the surrounding water, i.e., not as likely to generate streamers powerful enough to create a circuit.

The stepped leaders leave the cloud and have no particular target. Moving in only a general direction, they are pulled by the relative strengths of the fields below them. A rod will not easily shed a streamer, but when it does, the potential has built to the point that the streamer will be powerful. Such a streamer will rise a long way to reach the descending stepped leader. By contrast, a static dissipator with its array of sharp points will easily shed a charge and each point only exhibits the potential reached on it individually.

What we are trying to do is regulate the ionization process. We need to stay beneath whatever rate of discharge will attract the leaders coming from above. Our inability to determine exactly what this threshold will be means that the static discharge process can only be one weapon in our arsenal of lightning protection. We must acknowledge that the vessel may still take a strike. Actually it is relatively easy to build the same protection we would normally require in a rod system, and combine it with a static dissipator But now the normal protection system benefits from a reduced chance of a strike in the first place.

The Lightning Master Corporation of Clearwater FL. designs and manufactures systems for protection of broadcast towers, airports, and cellular phone installations. They have shown the effectiveness of the static discharge principle in deterring lightning strikes. Another company marketing dissipators is Island Technology of Cherry Hill New Jersey. Dissipators mounted on cellular telephone cell sites in the Tampa, Florida area dramatically reduced the number of sites damaged through direct lightning strikes. After a test, the dissipators have been installed on all of the towers in the area, as well as on many of the TV and radio broadcast antennas. Lightning Master makes a unit for sailboat masts which is available through Forespar Marine of Rancho Santa Margarita, CA. Information about the Lightning Master system can be obtained by calling Forespar at 714-858-8820 or by fax at 714-858-0505. The Island Technology units, called the No-strike System, also operate using the discharge principle, and work so well that the company will pay your insurance deductible up to $1,000 if a strike damages your boat. Island Technologies can be reached directly by calling 800-787-4535. Units are available for various size boats, but all feature an array of fine points to accept and disperse the static charge accumulating on your boat.

VI

PROTECTION DURING THE STRIKE

The stepped leader has reached an altitude of several hundred meters above the ground. A streamer has been formed and is moving skyward. They meet, and in two tenths of a second three to four flashes have transferred 25 coulombs from the cloud to the ground. The peak current flow may have reached 50 kilo amps.

The damage in this strike may be comparatively minor. The peak current never reached a high enough level for long enough to create a great deal of residual heat. But, it should be noted that this typical lightning strike may never actually exist. The range of current extends several orders of magnitude in either direction.

Our job is to ensure that whatever current reaches the boat can easily pass through it. In order to do this, we must give the current a path, and design the path so the electric current can follow it without being sidetracked.

There are a number of paths down from the masthead. All the halyards are potential conductors; as are the data lines running to the sensors, and the wires going to the lights. Even the walls of the mast itself can carry the current to ground. Our strategy must be to decide where we want the current to go, and make the resistance in that path as low as possible relative to the other possible routes.

The mast itself is generally the best route because it has relatively low resistance to electricity due to the wall thickness and cross sectional area. It also terminates in a location where the current can be directed off the boat. If the mast is wooden, a conductor of lower resistance than the shrouds must be added to it. All other possible points are firmly mounted to the masthead. This mounting can bond the other sensors, etc., to the mast electrically as well as physically. If one particular object is struck, the others are going to experience significant current through induction. We want this current to transfer to the mast by making all equipment on the mast share a common ground. Protecting the instruments themselves will be discussed later.

The characteristics of a good electrical bond are fairly simple. Direct metal to metal contact will suffice. It is most important to remove any corrosion coatings at any point where this contact occurs. All paint should be scrapped away, and even any anodizing removed. After the connection is made, the entire area can be sealed from the outside. The sealant will protect the area from atmospheric corrosion, and maintain a good bond.

It is important to remember through this stage that the electrical buildup we are working around is a surface, static charge. The charge forms by the elements on the outside, closest to the cloud charge, losing electrons. The charge collects on the outside surfaces, not through the internal wires carrying the current necessary to operate any of the devices, nor through the wires carrying the induced current of radio reception.

The coaxial cable used in antenna installations is usually shielded from this static build-up. The shielding is a woven conductor to designed to intercept any possible induced current from the outside, and stop it from interfering with the RF signal being carried by the antenna lead. The problem during a lightning strike will be the current carried by the shielding itself. The shielding will be able to carry current produced by the strike as well as carry the current from induction, so we must also plan for the disposition of this current when considering the protection of our instruments.

Since the mast acts as the transit to ground for the charge, we must next look at the base of the mast to see how the current will exit the boat. The base of the mast is a critical area to be considered in order to avoid catastrophic damage. The current will leave the mast somewhere around the base. If we can direct the current to a place of easy exit, it will pass merrily down into the limitless ground provided by the sea. However, if the current is stymied at the base of the mast and starts bouncing around on its own, it will leave the boat through whatever means are available; perhaps seacocks, through-hull fittings, speedometer transducer, or directly through the hull. The greater the difficulty the current has getting out, i.e., the greater the resistance of its chosen path, the greater the heat that will be generated. This heat is different than the heat generated in the bolt itself, but it is no less hazardous. If the current leaves through a non-metallic seacock, it will probably melt it. You need not worry about the possibility of extensive heat or fire damage though, because the hole left by the melted through-hull fitting will solve the problem by sinking the boat.

A pathway must be designed to guide the current from the mast to whatever is connected to ground. The characteristics of this pathway are straightforward. The pathway must be a conductive material of low resistance. The resistance must be low for the current to pass through the path without creating temperatures high enough to melt the conductor. It must lead to ground in the most direct path available. It must have some sort of terminal at the ground end so the charge can be transferred easily.

The proper conductor to bring the charge down the mast, or to carry it off the mast to the ground plate ideally would be a solid copper wire with a minimum diameter of 8 AWG. If the wire is multi-stranded, each strand should be 20 gauge wire. This diameter is needed for two reasons; to avoid the excessive heat build-up that will occur in smaller wire, and to avoid an overload of current that might encourage a corona-type discharge from the wire, which would result in the current electing some other path to ground.

Putting the connecting terminals on the mast will involve adding another type of metal to the mix there. We have an aluminum spar, a copper conducting wire, and a steel terminal. There will inevitably be some corrosion at the point of connection. This corrosion can easily defeat the entire purpose of the path. Just as with the masthead connections, this connection at the base has to be made carefully. Make sure the area is free of paint, and then carefully make the attachment. After the conductor is carefully secured, spray the entire assembly to limit corrosion.

The next stages of this system are the most difficult to get right. Part of the problem is that some strikes are not going to generate enough current to really test the system. The current will blithely follow the path like a well trained dog. A strike several orders of magnitude more powerful may not follow any path at all, and even the slightest resistance, or restriction may find it jumping off to another path.

Thus far we have simply taken the charge from the mast, straight down to the mast base and onto a conducting wire. The wire must lead to a discharge plate, or mechanism of some sort. This discharge plate provides a larger area for the electrical potential of the boat to contact the potential of the ground. It must be shaped to easily pass the current along. A rectangular shape has been found to be best. If the boat is equipped with a ham radio, the ground plate for the antenna system works well in this role.

This wiring system must provide the final route from the boat to ground. The suggested dimension for the ground plate is about one square foot. In the days of copper clad hulls, finding a ground was no problem. Modern yachts constructed of fiberglass may need to have a ground plate installed.

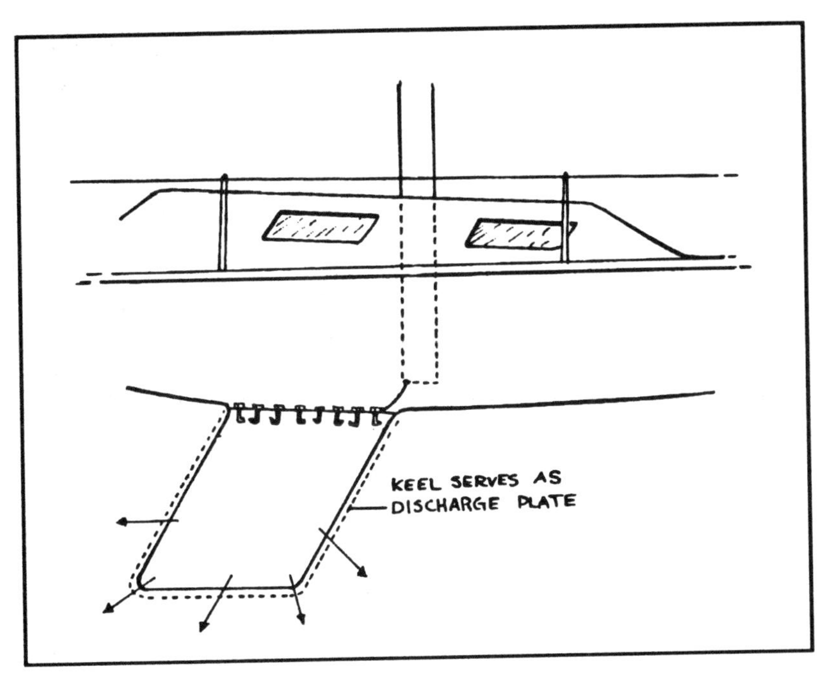

Figure 10. The current must be given as direct a path off the boat as possible. The conductor connecting the mast to the keel must be led in a smooth curve of at least 8 inches in radius.

Let's look at a couple of cases. A small cruising sailboat with a deck stepped spar, and a centerboard is the most difficult scenario. The current flows down the spar to a plate on the deck. The plate has to be bonded to the compression post underneath the spar. A conducting wire can pick up the current at the post, and take it where?

The compression post usually reaches the hull near the end of the centerboard trunk. A discharge plate should be mounted down each side of the trunk. Not only would this mounting provide a sharp edge for the trunk boundary, but the plate could be easily inset to be flush with the rest of the hull. The conducting wire can be led to the leading edge of the plate from the compression post without taking a sharp curve. The plate mounted along the trunk edges would also provide a place to mount flaps to smooth out the water flow around the base of the trunk while the board is down.

The keel stepped mast of a slightly larger cruising, or racing sailboat presents a similar problem. The conductor has to be led from the mast in a smooth curve of at least 8" radius. Typically, this wire is brought aft from the mast and mechanically attached to the keel. A lead keel is an excellent ground plate. The small amount of fairing putty and paint around it shouldn't be much of a barrier to the current reaching the sea.

Careful consideration has to be given to the path taken to reach the keel. The keel bolts are often in a sump underneath the cabin floor. If the ground conductor has to take two turns to reach a bolt, it's effectiveness may be impaired. It is better to drill a limber hole through a stringer to allow the ground cable to pass to the keel in one smooth curve. The NASA system employs a catenary for the same reason. The smooth curve from the top of the tower to the ground point allows the electricity a simple path. If the keel sump extends below the canoe body of the hull, and the bolt heads are on the bottom of the sump, it may be smarter to attach to the second or third bolt to keep the wire straighter. The grounding wire can be led down one side of the sump so it is not just hanging out in space. Then, the wire can be attached to a washer or a plate that is attached to the keel bolts.

Obviously, the low areas in the boat tend to collect water, which makes corrosion a critical factor. Eventually the copper wire, or a terminal will lose connection which will ruin the effectiveness of the entire system. Remember, the basis of the protection lies in the low resistance of the path we give the charge to follow. If corrosion impedes the current, or disrupts the continuity of the circuit, the current will find other ways to go. A ground wire hanging in the air does no good at all. It needs to be checked periodically.

If the yacht has an auxiliary engine, you can use the exposed surface of the shaft and prop to act as your ground plate. The continuity through the engine is usually sufficient to deliver the current to the exposed areas. Since the engine may be located significantly aft of the spar, the lead for the ground cable becomes important. The cable should again pass underneath the cabin sole, and must be protected against breaks and corrosion.

It must also be guarded against carrying any current caused by leakage from the other systems of the boat. The corrosion of the prop and other exposed parts will be hastened by a low level of current through them at all times.

Grounding through the engine is not as favorable as having a separate discharge plate. With engine grounding, the possibility always exists for damage to bearings and other parts through the massive heat which may accompany a particularly powerful or long duration strike. However, if a plate is installed, the engine should be grounded to it for reasons we will discuss when considering bonding.

Power yachts present a slightly different situation. Most of them do not have the ready path to the interior of the boat presented by the keel stepped mast. Any mast they do have may also not be significantly higher than other electrically attractive points on the superstructure. On the other hand, the hulls of power yachts are much more likely to be constructed of metal, aluminum or steel, which makes a ground plate superfluous.

Windage is less of a problem for power yachts. An "air terminal", an installation designed to attract a lightning strike, may be placed in a position to draw the strike away from less easily protected areas. This installation has to be high enough to cover the rest of the boat in the "cone" of protection. The air terminal must be the highest point on the yacht. A line drawn from the tip of the terminal 45 degrees below horizontal should cover the entire boat if the installation is sufficiently tall.

Power yacht grounding installations need to account for the additional areas of static charge accumulations. These can be corners on the flying bridge rails, or light masts. We can establish a common ground for these points, and then install a single air terminal, or a static dissipator. The establishment of the common ground will allow the dissipator to bleed off the excess charge, or place it in a position to take the strike. Since power boats don't have a dominant structure, the dissipator, or air terminal will have to be erected to handle the strike. A wrapped glass antenna, like a single side band, or loran installation, does not make a good lightning terminal. The antenna doesn't have the characteristics of being electrically sharp. It cannot bleed off the charge to deter the strike, and it wouldn't generate streamers to bring on the return flash.

On the other hand, the sharp corners of a flying bridge do have the capability of accumulating a charge and must be dealt with in precisely the same way we would want to ground a mast. We can make the same provision for a grounding conductor, and tie it into the main system. Since there is an easy path for the electrons to travel, any static charge which starts to build up on one part of the boat will equalize over the entire boat. Because the charge concentrates with height, the dissipator will either bleed the charge into the atmosphere, or receive the strike.

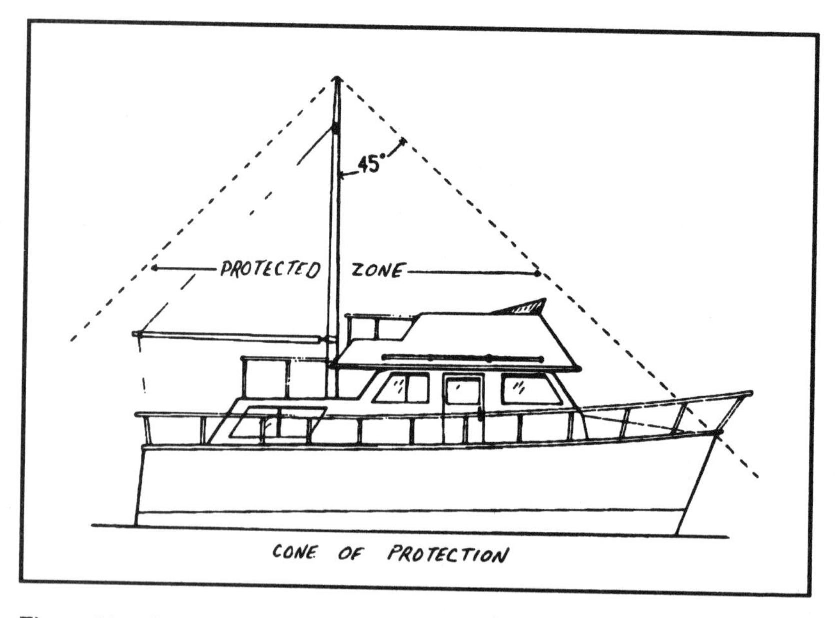

Figure 11. On powerboats, an air terminal will create a theoretical zone of protection at approximately 45 degrees from its tip.

The effects of a direct strike by a bolt of lightning are really completely unpredictable. We can set up our grounding system, have the proper wire, install a good discharge plate, and erect a proper terminal to receive the charge. But, we may get a positive stroke that carries so much current that the wire will melt, or we may have a corona discharge at a small bend, and the current may exit through a plastic seacock, melting it and sinking the boat. Nevertheless, the protection gained by having a system to take the flash may mean the difference between a natural event that occurs without consequences, and having to raise a sunken yacht. The chances of being hit by lightning are fairly small to begin with. We can further reduce the chance of significant damage by guarding against the larger percentage of strikes.

VII

THE SECONDARY EFFECTS OF LIGHTNING

As the return flash reaches the clouds, the electric potential between the cloud and the object struck is reduced. After the series of flashes run their course, the charged channel between the ground and the cloud breaks down, and current no longer flows. The current from the direct discharge has passed through, but the effects of the lightning are not limited to the simple exchange of electrical potential between the cloud and the ground. It is vitally important to be fully aware of the secondary effects of lightning strikes and consider them in the overall scheme of our protection system.

As the current in a strike finds its way to ground and neutralizes the potential difference that led to the strike, another difference of potential is created between the ground portion of the circuit and the area surrounding it. This potential must also equalize, and current will flow to the area of impact from the surrounding areas.

If this secondary current arcs, it becomes what are known as side-flashes. The potential develops the same way the initial field builds up that caused the strike in the first place. The charged field is dragged along underneath the opposite charge in the cloud. The charged field hasn't reached the stage where it can generate its own streamer, or it may not be close enough to the field in the cloud to draw the strike. Then, once the strike does occur, and passes through to the ground, the potential is neutralized for everything sharing the common ground. The areas not sharing the same ground will also seek to neutralize their potential versus the area struck. Current will flow, and may travel where it is not wanted.

Like any electrical current, these side-flashes will find the path of least resistance to go where they can equalize their potential. The danger is that the path of least resistance will lead through human bodies. Often a crew member will be in contact with two or more items of metal, say the aft rail and the helm. The potential built up by the static charge on those items will seek to move as electrical current. The current will flow toward the ground of the area struck. If the crew member is in that path, the current will flow through him. Particularly dangerous is if the upper part of the body is in contact and acts as a bridge, bringing current through the heart.

There is a story about cows that were standing around a tree struck by lightning. All the cows facing the tree were killed, while those facing away from the tree were left alive and unhurt. The answer to the mystery was that the cows facing the tree had an equalizing current pass through in a path which stopped their hearts.

Side-flashes also bring all the other hazards of unplanned, high powered current flowing through the boat. If the electrical systems pick up the current, the wiring may be damaged. The heat necessary to melt the wiring may cause other damage to the interior of the vessel. A power surge may travel through sensitive equipment and cause further damage.

Another secondary consequence of a lightning strike is the surge of electromagnetic energy known as the electromagnetic pulse (or EMP). This pulse may induce current in areas that would normally be completely isolated from the strike and the side-flashes. The primary danger from the electromagnetic pulse (EMP) is damage to the vessel's electronics. This damage can take many forms, such as: loss of internal memory, complete destruction of circuitry, or problems due to a temporary magnetic field. The issue of protecting electronic equipment will be covered in the next chapter.

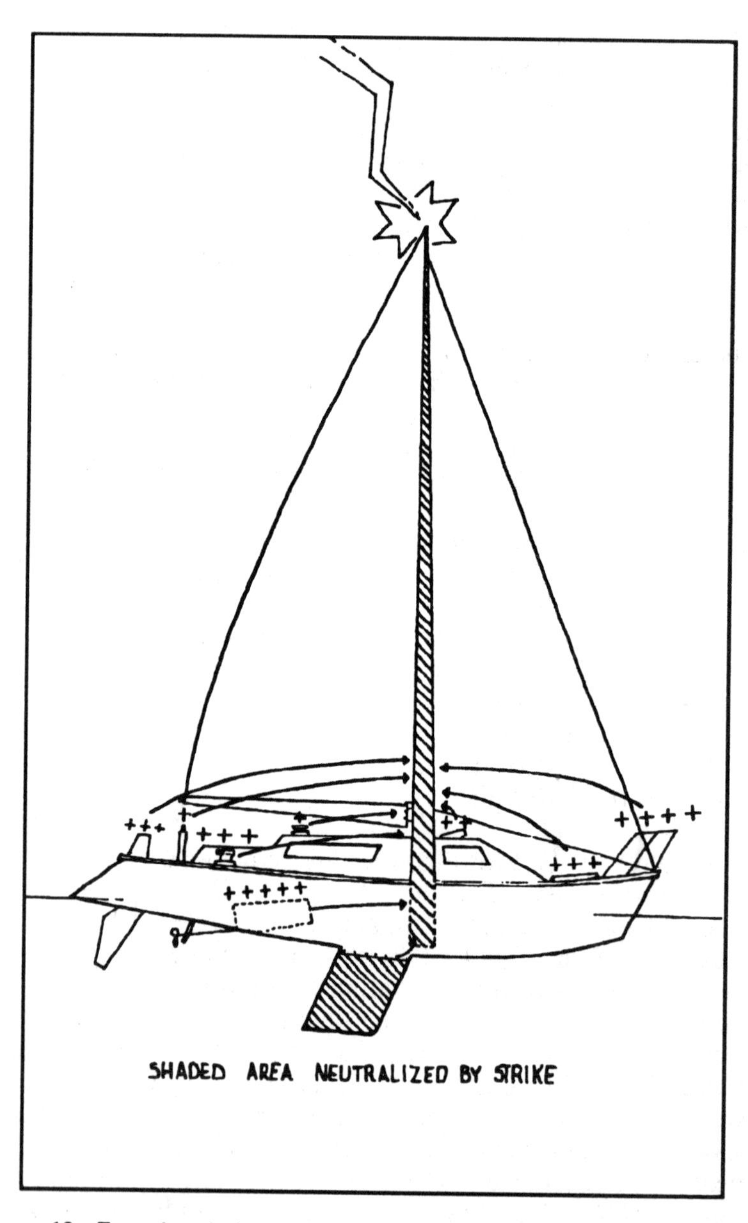

Figure 12. Even though the mast is grounded, the rest of the boat may retain a static charge. This potential will want to neutralize itself by flowing toward the ground. The resulting side-flashes can be as dangerous to the vessel and crew as the strike itself.

PREVENTING SIDE-FLASHES

The key to preventing these secondary current flows is to establish a common ground for the entire boat. Each major concentration of metal, or any area which might collect a large static charge must be grounded to the same discharge system used to protect the boat from the effects of the lightning strike itself.

This can be as simple as attaching a copper wire, again at least 8 AWG, from the item to be grounded, and leading it to the discharge terminal, or to a bus which in turn has a connection with the discharge plate. This system has to be outside of the normal ground installed for any electrical motors at the location.

The potential we are trying to neutralize is a surface charge carried by the casing, or the surface of the metal itself. The charge can be held by the metal rails, or it can reside in the casing of an anchor winch. The potential is going to travel through the air, hence the side-flashes, or it is going to make use of a bridge of opportunity - a human body, wet lines, or whatever. Giving the current a path to follow, a path of low resistance, will direct it where we want it to go - off the boat through the discharge plate.

The simplest method for this is to attach the wire lead to one of the mounting bolts for the equipment in question. The connection should be clean, and the same procedure followed as we used at the masthead. The connection should be protected from outside corrosion by coating the entire assembly. The ground wires can be led through the interior of the boat to the grounding terminal, or to a general bus terminal prior to the ground plate. The bus system may be preferable because it gives us a central point to ground the electrical system as well as control over the static charges on the boat. One of the interesting aspects of this secondary current flow is that it may happen even when the lightning strike does not score a direct hit on the boat. If the boat is moored, and the lightning strikes a tree nearby, the same discharges could occur between items on deck with differing ground potentials. By linking the areas of static charge on the boat to a common ground, we can prevent this discharge from taking place on the yacht itself. If we prevent this type of current from flowing around the boat, the current will also be deterred from damaging sensitive equipment, like the loran.

THE EMP (Electromagnetic Pulse)

The EMP causes problems by inducing electrical current in circuits as the pulse passes by. High school physics illustrates the link between electricity and magnetism. The pulse generated by a lightning strike is powerful, but brief. Most radio and electronic equipment is adequately shielded. The shielding absorbs the magnetic energy, causes it to convert into current, and disposes of the current through the ground in the instrument itself.

Adequate grounding of the instrument into the boat's electrical system should alleviate most of the effects of the pulse. The key consideration is establishing and maintaining a good connection between the chassis of the instrument and the common ground for the rest of the boat.

Most of the instruments are connected to their remote controls by some sort of coaxial cable. The cable has its own shielding to filter out normal interference from the signal. The shielding will pick up some current from any nearby pulse, especially a pulse as large as the one generated by a strike. This current has to be delivered to the chassis of the instrument and then off into the common grounding system.

The effect of the magnetic field passing through the boat is most pronounced on the compass. Fortunately, the effects are temporary. The compass will return to normal shortly. It is important to re-calibrate the deviation tables on a known course as soon as possible after the boat has been struck. If the strike occurs in transit, all possible opportunities to establish a range to check the accuracy of the compass should be utilized.

Any of the secondary effects of lightning can be just as destructive to the equipment on board. The low power requirements of the integrated circuit is a double edged sword.

Any power surge is going to affect its function. A sudden rise in temperature could render it useless. All this means is that to really protect the boat and its systems, we have to be more thorough in our efforts to prevent current from traveling around the boat. Although the most feared dangers from a strike are loss of life, serious injury, and catastrophic hull damage, the most common will be from secondary effects.

VIII

PROTECTING YOUR ELECTRONICS

For several reasons we've already mentioned, the most likely area of lightning damage is in the sensitive electronic components of your boat. The strike does not have to be on the boat itself to cause significant problems. The electromagnetic pulse from the lightning even in a cloud above your yacht can be enough to significantly screw up the works.

Remember that microprocessors work on extremely low voltages and can be easily "cooked" by even a modest power surge. A surge may merely disrupt whatever activity is in process, or it may remove all stored information from memory. In fact, if the current heats the system enough, it could render the entire system useless. These power surges are referred to as spikes.

The usual way to handle these hazards is to place protecting devices in line to isolate the sensitive circuits from these spikes. These surge protectors work by using materials that will only conduct currents above a certain baseline. When the surge comes through; the surge protector acts as a bypass, cutting off the circuit from the instrument before damage can occur.

Surge protectors work fine for most spikes, perhaps even those created by an EMP or a side-flash. However, current that has spanned 2-3 kilometers of air to get to the boat is usually beyond the capabilities of a surge protector. A strike is not going to be shunted to one side so easily. In fact, the actual current from a strike has got to be diverted using our primary ground system. If any electronic device is to survive, then the ground system we have installed or arranged has to work. Next, we can work on saving the sensitive gear from the effects of the EMP and secondary current.

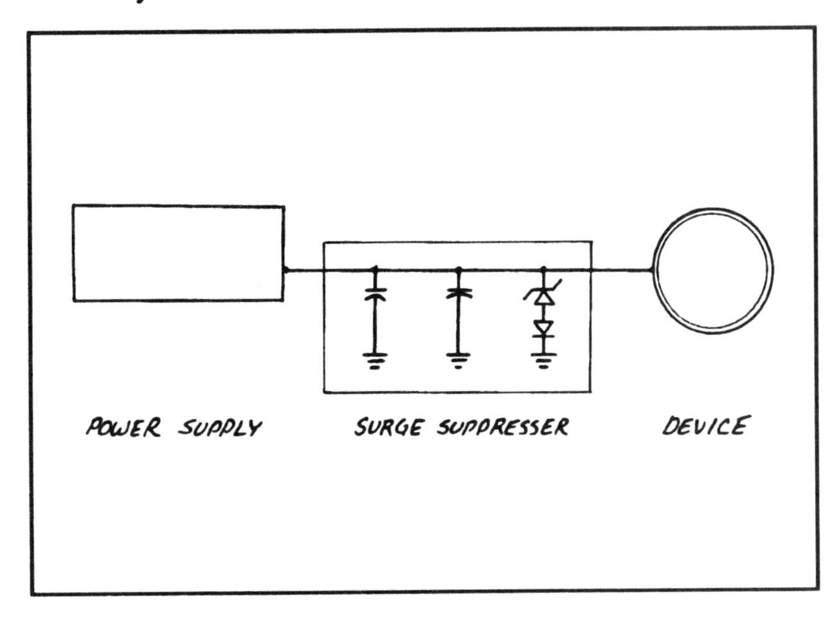

Figure 13. Electrical diagram of a surge protector.

Some of the protective measures run counter to what we might intuitively think. Take the case of our masthead instruments. A widely used wind sensor incorporates a chip in the unit itself. My common sense says that to protect this chip, we can surround it with an insulator to prevent a charge from reaching it. This was apparently the manufacturer's thinking as well, as the early models were encased in a non-conductive housing. After the units were coming back damaged from various electrical problems, a reversal of thinking occurred.

If the casing is conductive and has a lower resistance than the interior circuits, the casing will guide the current around the interior circuits. If we also provide the casing a ground of its own, the current will probably pass around the guts of the instrument without catastrophic consequences.

Herein lies the entire key to effective protection. The charge that we deal with is primarily one of static, or surface electricity. Even the induced current produced by the electromagnetic pulse is going to be primarily evident in the chassis and shielding of the various pieces of equipment and their assorted power and data cables. We need to give this static current a clear path to a ground. The proliferation of electronic devices, and their remote sensors and readouts makes the problem a little more difficult.

Eliminating the difference in ground potential is again the major issue. All items that have the ability of carrying current, whether it is from the basic power supply of the yacht, or the result of potential differences built up from static sources, must be linked into the primary ground system.

The area at the top of the mast is again the most vulnerable. The wind sensors and other instruments mounted on the top of the mast represent an area of significant static concentration. As we have seen, the buildup of potential can be handled by giving the current developed by the potential an easy way down the spar. Bonding the instrument to the mast itself is the answer. Most threaded attachments are mounted using some substance to prevent the threads from backing out. This can be a danger because the same material is going to cause resistance when the current tries to pass onto the mast. What is needed is a good electrical connection to prevent any of the current flow from taking an alternate route. If some kind of adhesive has to be used to mount the instrument, some other kind of provision must be made for a clean connection. A large washer may provide the necessary contact. If this is the case the connection should be protected from the outside to prevent corrosion from cutting the electrical connection.

The current created by the static charge (or side-flash) also moves on the outside of things. The shielding in coaxial cable may pick up the charge and move it down to the chassis of the instruments. We have to insure that the connection through the shielding is intact and delivers the charge to ground. If possible, we want to keep the charge out of the lines directly feeding the processing circuitry.

The shielding may also pick up current from the EMP. This current will be carried by the shielding on the coaxial cable and by whatever RF (radio frequency) shielding is installed over the chassis of radio receivers (loran, GPS, SSB etc.) and radar sets.

The solution to the problem is to link the shielding for the cables and the electronics, and the fixed metal chassis which holds the electronics, to the common ground shared by the rest of the boat. With this level of protection, lightning could strike the boat next door and no current would flow between items on your boat because they all share the same ground. The current generated by a side-flash and the electromagnetic pulse would simply travel to the ground, and off the boat.

Another area that needs some planning is in the chassis ground of the systems themselves. The chassis is the actual box and mountings that holds the circuit boards of a piece of electronic equipment. Remember that the problem is the rush of the static charge on the rest of the boat to fill the potential neutralized by the strike. This static charge is a surface charge, and moves on the outside of things. In order to keep this static charge out of the electronics inside the chassis, we must provide the charge with a means of exit to a ground. In this case, the entire instrument casing has to be grounded. Just grounding the power supply to the instrument is not enough to avoid damage from the stray voltage.

Cellular telephone installations use what is simply called a halo ground for this purpose. This is simply a low resistance cable that is routed around each instrument. A connection is made between each installation and the ground wire, and then to a single ground. The existence of the common link maintains the common ground all through the system. The common ground is again the key. Since instruments are generally concentrated in one area of the boat, this system is not at all difficult to establish. The main electrical panel is usually located nearby, and provides the ideal location for the bus to the primary ground.

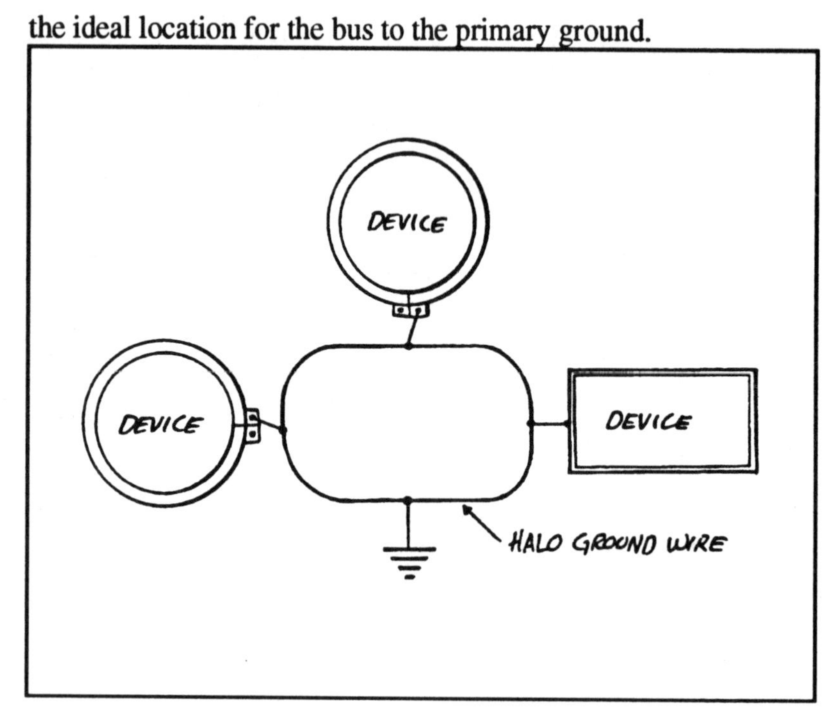

Figure 14. The halo ground. This system connects the chassis of each electronic device to the common ground.

If the lightning strike came aboard your boat, all the transfer of potential could theoretically take place through the ground wires. If the ground wires provide the path of easiest egress from the boat, then the current would dutifully flow off into the sea.

Many of the common preventive measures taken to protect the instruments will have only slight benefits. Often, people will disconnect antenna leads, turn off equipment, or make some other attempt to isolate their gear from the current generated by a strike. Experience shows that these attempts are mostly window dressing. The current flowing around the boat is not the result of a surge in the power supply, so turning off the power is not the answer. Most antennas are poor lightning attracters so disconnecting the antenna leads may help restrict the current from the strike, but will do little to offset the secondary effects which are responsible for most of the damage.

The single most important measure to prevent that damage is to tie all systems into the general ground. Bonding the chassis, and the power and data lines to the primary ground will at least remove the ground potential difference as a cause of current flow. If we eliminate current, we eliminate heat and arcing as well.

The final step to completing the bonding process is to bring the ground for the general electrical system into our lightning system.

IX

BONDING YOUR BOAT'S ELECTRICAL SYSTEM

The potential difference on your boat will try to equalize using whatever pathway is available. If the static charge is strong enough, it will go through moisture, bodies, or the circuits of the electrical system, and utilize the power lines within the boat to travel toward the opposite charge. We can prevent this from happening by connecting the ground from the electrical system to the same ground we use to disperse the electrical power of the strike itself.

This process is known as bonding the electrical system, and it simply means giving every possible collector of electric potential access to the same ground. As we have seen, the source of some of this potential is the static charge created by the proximity of a thunderstorm. While this is primarily a surface charge, it could gain access to the wiring if this would lead to a different ground potential.

We have previously recommended the installation of a ground plate to act as a ground terminal for the system designed to take the current from an actual strike. This terminal can also serve as the ground plate for the negative end of the boat's electrical system. Most systems are grounded to a common terminal which is then connected to the engine. The engine acts as a huge electron sink, and can pass the charge through the boat utilizing the shaft and propeller. This gives the electrical system a common ground and provides for the proper operation of the onboard systems.

We can either further ground the engine to the discharge plate, or we can connect the negative bus directly to the plate, and bring the engine into the circuit separately. Since the engine may gather a large charge of its own, it is important that the engine share in the common ground. Otherwise the engine would continue to ground itself through the shaft, and develop a different potential than the rest of the boat, which might try to equalize in some unpleasant way.

Most twelve volt systems have a common ground wire leading from the vessel's electrical panel to whatever is currently being used as the ground. It is a relatively simple matter to lead that cable to the ground plate. The cable will have a short run, and we can use a large diameter cable to carry the load. This large cable will have a correspondingly low resistance, and will easily transport any current to the main ground potential. No matter where the current comes from, this last bit of cable will be carrying it to ground, and must be capable of handling a large current without melting.

The link between the engine and the ground plate also has to have an extremely low resistance. One suggestion for this connection is to hammer a piece of copper tubing flat, and use it to bridge the gap between the engine and the plate. This would certainly make the path of least resistance lead to the plate rather than through the shaft and propeller.

In a similar fashion, every fitting through the hull should be connected to the ground plate. Unless the fitting is plastic, and totally isolated from any electrical activity, it should be bonded into the system.

Tying the electrical system of the boat to the outside like this causes some problems. The incursion of stray current from the outside speeds the process of corrosion. If stray current is present, zincs mounted as sacrificial anodes will corrode away much more quickly than normal. Stray current is usually the result of being next to an improperly grounded shore power installation, and is fairly common inside large marinas.

Beyond the zinc problem, the wood through which a seacock may be mounted will decompose more rapidly if it is subjected to galvanic current. The first indication will be some discoloration surrounding the fitting itself. The hazard here is that the wood completely rots away, and destroys the integrity of the fitting. If this starts to occur, the solution is to either drop that fitting from the circuit, or replace a metallic fitting with a plastic one.

A well bonded boat should be able to take the sudden potential shift accompanying the strike, and neutralize it across the entire boat through the grounding system. Without a potential difference developing, the occurrence of side-flashes or secondary current flows through wiring should not occur.

If the boat suffers an extremely powerful strike, or encounters a positive strike, even the best system may be overcome. The potential moved around the boat may be so great that the ground connections may overload and melt. The bonding system by necessity is going to have more curves than the primary ground lead, and the secondary flows may produce enough current to jump off at those points. These corona discharges have the capability to scramble the electronics as easily as the direct current of the strike.

The point is that it is extremely rare to be hit by lightning. It is rarer still to be hit with a massively powerful strike. But, it is not so rare to be located next to something that is struck. Careful attention to bonding will give the equipment a much better chance of suffering no ill effects from the event.

Part of the problem with a passive system like this is that there is a tendency to set it up and forget it. Discussed in the next chapter are some areas that should be looked over in your spring commissioning.

X

THE SYSTEM AT WORK

We have now put in place the three components of our protection system. Our first line of defense is the static dissipator. As the charge drawn under the thunderstorm moves across the terrain, the dissipator allows the charge to stream into the atmosphere at a low level of field strength. Ideally this field strength will be below that of surrounding objects and a lightning strike will be drawn elsewhere. Should the dissipator still draw the strike, the current flow is directed through the boat to a discharge plate which acts as the ground.

Once a strike triggers an exchange of potential, the task of our system is to deliver the current through the boat to the ground. Our system carries the massive current load by providing the path of least resistance through the boat. By using large diameter cable, or the walls of the spar, we direct the current to the discharge plate. Remember that the path to the plate from the point of impact must be as direct as we can make it. Any tight curves, or abrupt changes in direction may give the current an opportunity to jump off the wire, and take its own path to the ground.

Once the current reaches the discharge plate, it should flow to the limitless ground represented by the water. The plate has got to be sufficient in size to distribute the current. It should have sharp corners and edges to give the current a good jumping off point.

As the current passes through the boat, it neutralizes the potential which had attracted the strike. The bonding of the major metallic components on deck, and the grounds of the power supply allow the electric potential that has accumulated throughout the boat to also neutralize itself through the same process. Low resistance pathways through the boat leading to the ground plate draw the current off without it jumping through places where it can do harm.

The existence of these three safeguards; the dissipator, the ground system, and electrical bonding, reduce the chance of lightning damage. Unfortunately, the active word is still "reduce." Studies done on the cellular phone system have proven that these methods are effective in significantly reducing downtime associated with lightning activity. They do not, and cannot eliminate the danger of a lightning strike. What they do is to take an already unlikely event, and make the possibility of damage even more remote.

MAINTAINING THE SYSTEM

One of the nice features of this three-tier lightning protection system is that there are no moving parts. It is a passive system. However, it is not maintenance free.

The proper functioning of the system depends upon the ground pathways maintaining a low level of resistance. Corrosion of the points of attachment is always going to present a problem. It is just like blocking the water flow from a hose. The water squirts out somewhere. The whole system depends upon the current going where we direct it. All connections must be electrically sound. This will take some inspection, and some cleaning if necessary.

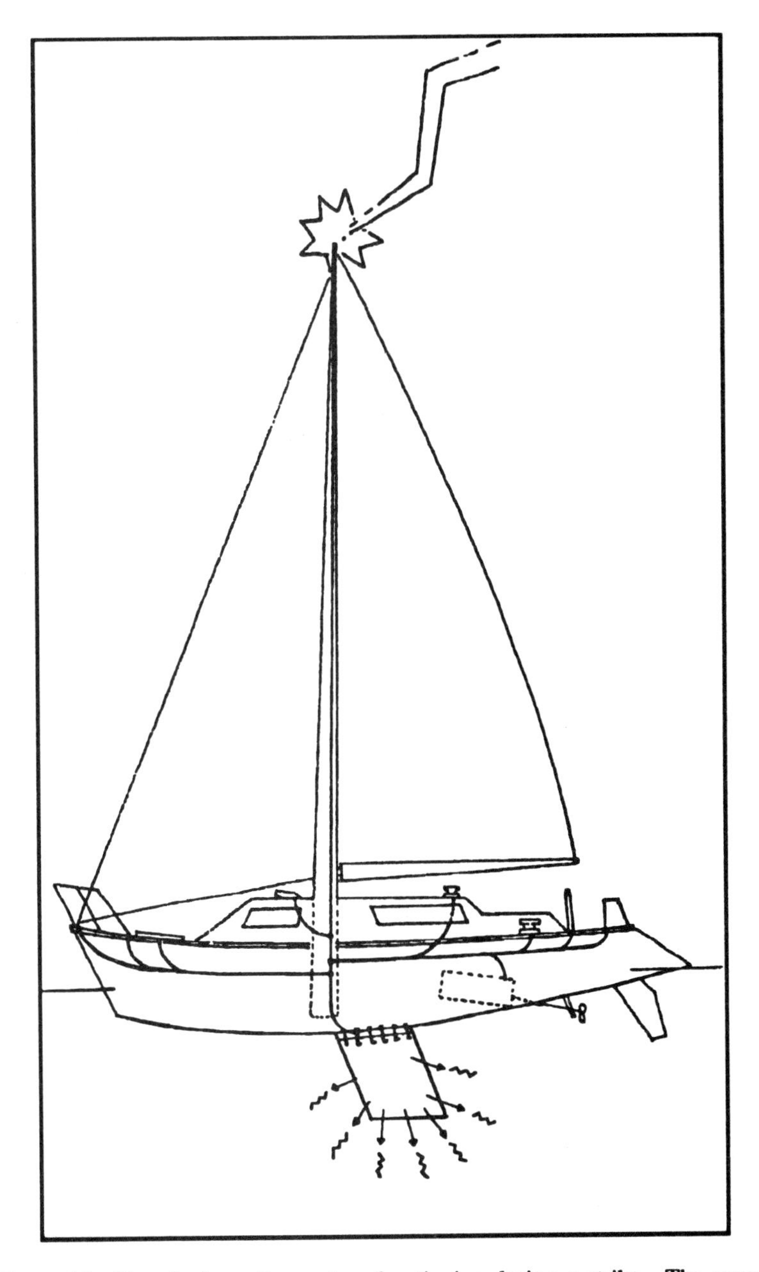

Figure 15. Here is the entire system functioning during a strike. The current from the strike is carried through the boat to the ground. All other static concentrations flow along the bonding connections and neutralize through the same ground.

One area to be watched is the wire lead coming from the mast base to the keel bolts. The bilge of a boat gathers water, and corrosion can be a problem. I have lifted the floorboard of a J-24, and found the ground wire hanging in the air. The connection had completely corroded through. Any current from a strike would have probably proceeded down the mast, and blown out of the boat through the speedometer fitting. The resulting hole could have quickly sunk the boat.

The bonding process also requires some monitoring. We need to be on the lookout for both the electrical integrity of the system, and the existence of stray current, and its galvanic consequences. As we've mentioned, the wood mounting blocks for various fittings will turn dark in the presence of this type of activity. When galvanic current is suspected to be causing a problem at a fitting, it might be a good idea to remove that item from the circuit and monitor the progress of the discoloration. This can be done after testing for stray current emanating from the electrical system. If stray current is found, then tracking down the current leak would certainly be a priority.

Lightning is a common phenomenon. Being struck by lightning is not so commonplace. Unfortunately that does not help deal with the catastrophic damage associated with the event. Following the systems outlined in this work will not guarantee that you can escape the damage, but it will increase the odds in your favor.

The Galvanic Series for Metals in Salt Water

Magnesium and its alloys
Zinc
Aluminum and its alloys
Cast Iron, Mild Steel
304, 410 Stainless Steel
316 Stainless Steel
Aluminum Bronze
Brass
Tin
Copper
Manganese Bronze
Silicon Bronze
Gunmetal
410 Stainless Steel (passive)
400 Stainless Steel (passive)
90-10 Cupro-nickel Alloy
430 Stainless Steel (passive)
Lead
70-10 Cupro-nickel Alloy
Nickel Aluminum Bronze
302, 304, 321, 347 Stainless Steel (passive)
Monel 400, K5OO
316, 317 Stainless Steel (passive)
Titanium
Graphite, Carbon Fiber

When two metals are connected, the one higher on the chart will corrode. The greater the distance in the table between the two metals, the greater the potential difference and the faster the corrosion.

REFERENCE MATERIAL

BOOKS

Calder, Nigel. 1990. Boatowners Mechanical and Electrical Manual: How to maintain, repair, and improve your boat's essential systems. Camden, Me.: International Marine Publications.

Maloney, Elbert S. 1989. Chapman's Piloting, Seamanship & Small boat Handling. 59th ed., New York, NY.: Hearst Marine Books.

Uman, Martin A. 1984. Lightning. 2nd ed., New York, N.Y.: Dover Publications

PAMPHLETS

Becker, William J. 1985. Boating - Lightning Protection. Gainesville, Fl.: Sea Grant Extension Program, University of Florida.

Kaiser, Bruce. 1990. Lightning Protection for Boats: A Three-Pronged Attack. Clearwater, Fl.: Lightning Master Corp.

NASA. 1990. Lightning and the Space Program. Release 72-90. Kennedy Space Center, Fl.: National Aeronautics and Space Administration.

GLOSSARY

Bonding, Electrical: To establish an electrically conductive connection between two items. To connect a number of items to a common electrical ground.

Catenary: The curve assumed by a flexible cord hanging freely from two fixed points. A cable suspended between two points.

Chassis: The frame upon which is mounted the working parts.

Convection: The process of heating and cooling which causes air to rise and new cooler air to be drawn in underneath. As the heated air rises, it cools and eventually descends to be heated again.

Corona Discharge: A discharge resulting from the electric breakdown of the air or other insulating materials surrounding a current of high voltage.

Cumulus Cloud: A large cloud usually forming between 2,000 and 15,000 feet above ground level. A cumulus cloud characteristically has a flat bottom and rounded top. Cumulus clouds forming by convection on warm afternoons and can be a precursor to a thunderstorm.

Discharge: The equalization of a difference in electric potential by a transfer of electricity. The character of the discharge is determined by the medium through which it occurs, the amount of potential equalized, and by the form of the terminal conductors on which the difference existed.

Electric Field: A region of space established by the proximity of an electric charge, or under the influence of an electrical charge. The resulting magnetic force caused by the introduction of an electrical charge on an object or in a region of space.

EMP (Electromagnetic Pulse): A temporary burst of electromagnetic radiation produced by the sudden initiation and cessation of an electrical current.

Graupel: Solid water droplets not yet fallen as rain.

Ground: A large conducting body used as the common return for an electrical circuit and as arbitrary zero potential. The electrical connection with the earth.

Halo Ground: A grounding system utilizing a single ground connected in series to the negative pole of a number of circuits, or electrically bonding a number of circuits together and to ground.

Leaders, Stepped: The paths followed by the ions in a thunderstorm prior to the equalization of potential. Stepped leaders precede the initial return flash, as compared to the simple leaders which precede all other return flashes in a lightning strike.

Potential Difference: The difference in electric potential between two points that represents the amount of work involved or the energy released in the transfer of one unit from one point to the other.

Resistance: The property of a substance whereby it opposes or limits the passage of an electrical current.

Return Flash: The visible portion of a lightning strike. The potential difference is reduced by the return flash following the path of ionization created by the leaders preceding the flash. The return flash emanates from the object struck.

Side Flash: A disruptive discharge between a conductor traversed by lightning or other high current, and neighboring masses of metal or between different parts of the same conductor.

Static Dissipator: An instrument designed to reduce the concentration of static charge upon any one area.

Streamers: An area of ionization extending from an area of strong static charge toward the leaders emanating from a thunderstorm. After a streamer meets a leader, a conductive path has been formed for a return flash.

NOTES

ABOUT THE AUTHOR

Michael V. Huck, Jr. brings a wide variety of sailing experience to bear on the problem of lightning and boats. He started racing scows at the age of ten and competed throughout the Midwest during his school career. After receiving his B.A. from Knox College in Galesburg, Ilinois, he spent three years in Europe aboard the Quest, a 176 foot tallship operated by the Flint School. He served as watch captain and first officer. After returning to the United States, he worked for Melges Boat Works in Zenda, Wisconsin, continued racing scows, and began contributing articles to the sailing press. Leaving Melges in 1986, Michael moved to Florida where he works as a freelance writer, and does a lot of sailing.